Gail and Jay were up early.
They were going to the fair!
"This will be a perfect day," said Gail.

1

Dad paid and they went in.

"Wait!" said Dad. "You must learn to stay with me."

Gail and Jay rode the train.

"Look this way!" said Dad.

"Smile and say bees!"

"May I play this game?"
asked Gail.

"Don't fail!" said Jay.

They heard people say,
"Step right up! See the best
ride on earth!"

They sailed into the air. The basket swayed.

"Look!" said Jay. "I see the earth!"

After the ride, they ate snacks. Gail and Dad ate from a tray. A man painted Jay.

They stayed until the end.
"What fun!" said Gail to
Jay. "It was a perfect day."

The End

Understanding the Story

Questions are to be read aloud by a teacher or parent.

1. Why did Gail and Jay get up early that day?
2. What are some of the things they did at the fair?
3. What did Jay say he saw from up in the air?
4. Do you think Gail and Jay liked the fair? Why do you think that?

Answers: 1. because they were going to the fair 2. Possible answers: rode the train, played a game, went in a hot-air balloon, ate, had faces painted 3. the earth 4. yes, because they said it was a perfect day

Saxon Publishers, Inc.
Editorial: Barbara Place, Julie Webster, Grey Allman, Elisha Mayer
Production: Angela Johnson, Carrie Brown, Cristi Henderson

Brown Publishing Network, Inc.
Editorial: Marie Brown, Gale Clifford, Maryann Dobeck
Art/Design: Trelawney Goodell, Camille Venti, Sarah-Beth Zoto
Production: Joseph Hinckley

© Saxon Publishers, Inc., and Lorna Simmons

All rights reserved. No part of the material protected by this copyright may be reproduced or utilized in any form or by any means, in whole or in part, without permission in writing from the copyright owner. Requests for permission should be mailed to: Copyright Permissions, Harcourt Achieve Inc., P.O. Box 27010, Austin, Texas 78755.

Published by Harcourt Achieve Inc.

Saxon is a trademark of Harcourt Achieve Inc.

Printed in the United States of America
ISBN: 1-56577-986-X

4 5 6 7 8 546 12 11 10 09 08 07

Phonetic Concepts Practiced

ai (train)
ay (day)

Nondecodable Sight Words Introduced

early
earth
heard
learn
were

ISBN 1-56577-986-X

Grade 1, Decodable Reader 24
First used in Lesson 69

Roy's Best Toy

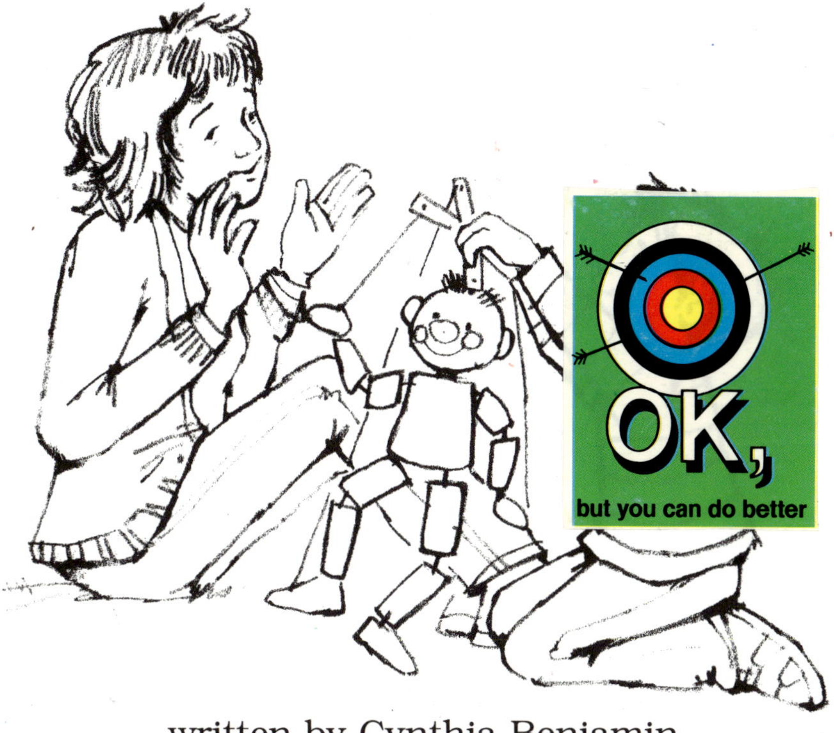

written by Cynthia Benjamin
illustrated by Taia Morley

SAXON PUBLISHERS

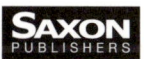

THIS BOOK IS THE PROPERTY OF:

STATE _____	Book No. _____
PROVINCE _____	Enter information
COUNTY _____	in spaces
PARISH _____	to the left as
SCHOOL DISTRICT _____	instructed
OTHER _____	

ISSUED TO	Year Used	CONDITION	
		ISSUED	RETURNED

PUPILS to whom this textbook is issued must not write on any page or mark any part of it in any way, consumable textbooks excepted.

1. Teachers should see that the pupil's name is clearly written in ink in the spaces above in every book issued.
2. The following terms should be used in recording the condition of the book: New; Good; Fair; Poor; Bad.